Abdelhafid Mimouni

Inguinal Hernia Bioinorganic Chemistry

Abdelhafid Mimouni

Inguinal Hernia Bioinorganic Chemistry

ScienciaScripts

Imprint

Any brand names and product names mentioned in this book are subject to trademark, brand or patent protection and are trademarks or registered trademarks of their respective holders. The use of brand names, product names, common names, trade names, product descriptions etc. even without a particular marking in this work is in no way to be construed to mean that such names may be regarded as unrestricted in respect of trademark and brand protection legislation and could thus be used by anyone.

Cover image: www.ingimage.com

This book is a translation from the original published under ISBN 978-620-6-71422-4.

Publisher:
Sciencia Scripts
is a trademark of
Dodo Books Indian Ocean Ltd. and OmniScriptum S.R.L publishing group

120 High Road, East Finchley, London, N2 9ED, United Kingdom
Str. Armeneasca 28/1, office 1, Chisinau MD-2012, Republic of Moldova, Europe
Printed at: see last page
ISBN: 978-620-7-68400-7

BIOINORGANIC CHEMISTRY OF INGUINAL HERNIA

AUTHOR

Dr Abdelhafid Mimouni is an independent researcher specialising in bioinorganic systems chemistry, with extensive expertise in macromolecular synthesis and characterisation. He obtained his PhD in chemistry from the University of Paris XII in 1997 and a Diplôme d'Études Approfondies in bioinorganic systems from the University of Paris XI in 1993.

SUMMARY

This book explores in depth the mechanisms of inguinal hernia from the perspectives of bioinorganic chemistry and molecular biology. It details the complex anatomy of the abdominal wall, highlights the interactions of metals and nutrients in tissue healing, and highlights the importance of trace elements and metalloproteins in the regeneration of muscle and connective tissue. It also examines recent scientific advances, including the potential impact of gamma radiation on the adrenal glands. This book is aimed at researchers, clinicians and students interested in these interdisciplinary fields, with the aim of enriching understanding of the biochemical and molecular aspects of inguinal hernia and stimulating interest in future developments in this crucial area of scientific research.

PLAN

INTRODUCTION

Definition and background

Inguinal hernia is a common medical condition that occurs when the contents of the abdominal cavity, such as part of the intestine, protrude through a weak point in the abdominal wall in the groin area. This condition mainly affects men, but can also occur in women and children. Typical symptoms include pain or discomfort in the groin area, particularly during physical exertion, coughing or lifting heavy weights.The causes of inguinal hernia are varied and include genetic factors, ageing, strenuous exercise, obesity, pregnancy, poor diet and conditions causing chronic coughing. These factors can weaken the muscles of the abdominal wall, making the groin area more susceptible to protrusion of internal organs. The incidence of inguinal hernias is significant, accounting for around 75% of all abdominal hernias. Approximately 27% of men and 3% of women will develop an inguinal hernia during their lifetime. Traditionally, the most common treatment for this condition has been surgery to repair the abdominal wall and replace the internal organs. However, like any surgical procedure, this option carries risks, such as infections, anaesthetic complications and recurrence of the hernia. In this book, we explore an alternative, non-surgical approach to the treatment of inguinal hernia. By adopting methods based on specific dietary changes and the application of a special hernia bandage, it is possible to promote the natural healing of the hernia. This approach is based on the principle that, just like a torn ligament or a muscle injury, an inguinal hernia can heal on its own if the right conditions are met.The importance of a non-surgical approach lies in its many potential benefits. It minimises the risks associated with surgery, avoids prolonged recovery periods and enables patients to return to their daily activities more quickly. In addition, this non-invasive method can be particularly beneficial for individuals with contraindications to surgery or who prefer to avoid surgery. By following the detailed instructions in this book, patients can take control of their inguinal hernia treatment and promote a natural and effective healing process.

CHAPTER 1

ANATOMY AND PHYSIOLOGY OF THE ABDOMINAL WALL

Anatomy of the Muscles of the Abdominal Wall

The abdominal wall is a complex structure made up of several layers of muscle, each of which plays a crucial role in protecting the internal organs, particularly the intestines. In the context of inguinal hernia, the internal abdominal oblique muscle is particularly important, although other muscles also contribute to this function. Understanding the anatomy and physiology of these muscles is essential to understanding the mechanisms of inguinal hernia and the potential reasons for their weakness.

Internal oblique muscle :

· Location: The internal oblique muscle is located below the external oblique muscle and above the transverse abdominis muscle. It extends laterally from the iliac crest and the inguinal ligaments, inserting onto the lower ribs and the linea alba of the abdomen.

· Thickness: The thickness of the internal oblique muscle varies according to the individual and their level of fitness, generally between 2 and 5 mm. This thickness can influence the muscle's strength and resistance to intra-abdominal pressure.

· Length: The internal oblique muscle extends from the bottom of the rib cage to the pubic bone, covering a length of 15 to 20 cm. This wide coverage supports and stabilises a large part of the abdominal wall.

Transverse abdominal muscle :

· Location: The transverse abdominis muscle is the deepest of the abdominal muscles, located below the internal oblique muscle. It extends horizontally around the abdomen, from the ribcage to the pubic bone, passing through the spine.

· Thickness: The thickness of the transverse abdominis muscle varies, but on average it is between 1 and 3 mm. Its position and orientation make it particularly effective in compressing the abdomen and stabilising the trunk.

· Length: Similar to the internal oblique muscle, the transverse abdominis muscle extends from the rib cage to the pubic bone, providing complete support to the abdominal wall.

External oblique muscle :

· Location: The external oblique muscle is the most superficial of the abdominal muscles, overlapping the internal oblique muscle. It extends from the lower ribs to the iliac crest and pubic bone, forming a large protective layer.

· Thickness: The thickness of the external oblique muscle is generally between 2 and 4 mm. Although superficial, it plays a crucial role in torsion and lateral flexion of the trunk.

· Length: The external oblique muscle extends from the lower ribs to the iliac crest and pubic bone, covering a large area and contributing to the stability of the abdominal wall.

The role of muscles in protecting the intestines

These muscles work in synergy to form a protective wall around the abdominal cavity, holding the internal organs in place and allowing movements such as bending, twisting and stabilising the trunk. The coordination and strength of these muscles are essential in preventing inguinal hernia formation by keeping intra-abdominal pressure under control.

Causes of Muscle Weakness

1. Genetic factors :

· Some individuals may have a genetic predisposition to a weaker abdominal wall, increasing their risk of developing a hernia.

2. Ageing :

· With age, muscles can lose their elasticity and strength, making the abdominal wall more susceptible to hernias.

3. Intense Physical Effort :

· Activities involving intense or sudden physical effort, such as lifting heavy weights, can increase intra-abdominal pressure, contributing to the formation of a hernia.

4. Obesity :

· Excess weight can put extra pressure on the abdominal wall, weakening the muscles and increasing the risk of hernia.

5. Pregnancy:

· In women, pregnancy can stretch the abdominal muscles and increase intra-abdominal pressure, which can weaken the abdominal wall.

6. Poor power supply :

· A diet low in essential nutrients can affect muscle health, reducing the muscles' ability to maintain and repair themselves.

Poor Nutrition and Muscle Health Importance of Nutrition for Muscle Health

A balanced diet rich in essential nutrients is crucial for maintaining the health of the abdominal wall muscles. Muscles require a range of vitamins, minerals, amino acids and proteins to grow, repair and function properly. A lack of these nutrients can weaken muscles, making them more susceptible to injury and conditions such as inguinal hernias.

Essential Nutrients for Muscle Health

1. Protein :

· Role: Proteins are the basic building blocks of muscle. They are necessary for muscle repair and growth, and for the production of new muscle fibres.

-Food   sources: Lean meat, fish, eggs, dairy products, legumes, nuts and seeds.

2. Essential Amino Acids :

· Role: Amino acids are the building blocks of proteins. Some amino acids, known as essential amino acids, cannot be synthesised by the body and must be obtained from the diet.

· Main Amino Acids :

· Leucine: Stimulates muscle protein synthesis.

· Isoleucine: Helps muscle recovery.

· Valine: Helps energy and muscle recovery.

· Food sources: Meat, fish, eggs, dairy products, soya, beans, lentils.

3. Vitamins :

· Vitamin D :

· Role: Essential for healthy bones and muscles, it helps with calcium absorption and muscle function.

· Dietary sources: Exposure to sunlight, oily fish, beef liver, cheese, egg yolks.

· Vitamin C :

· Role: Contributes to the formation of collagen, necessary for the repair of muscle tissue.

· Food sources: citrus fruits, strawberries, peppers, broccoli, kiwis.

· Vitamin E :

· Role: Antioxidant that protects muscle cells from damage.

· Food sources: Nuts, seeds, vegetable oils, spinach, broccoli.

· Vitamin B12 :

· Role: Important for the production of red blood cells and nerve health, it helps muscle regeneration.

· Food sources: Meat, fish, eggs, dairy products, cereals enriched.

4. Minerals :

· Calcium :

· Role: Essential for muscle contraction and bone health. Minerals:

· Calcium :

· Role: Essential for muscle contraction and bone health.

· Food sources: Dairy products, sardines, tofu, broccoli, almonds.

· Magnesium :

· Role: Helps muscle relaxation and energy metabolism.

· Food sources: Nuts, seeds, legumes, spinach, avocados.

· Potassium :

· Role: Helps maintain water and electrolyte balance i n   muscle cells.

· Food sources: Bananas, potatoes, spinach, avocados, beans. Impact of Nutrition on Tissue Repair

Maintaining an adequate intake of these nutrients is crucial for the repair and regeneration of muscle and connective tissue, which is essential for the healing of inguinal hernias. In particular, vitamin C is necessary for collagen synthesis, which is important for tissue strength and elasticity. Protein and essential amino acids are necessary for the growth and repair of weakened muscles. Antioxidant vitamins and minerals, such as vitamin E, protect cells from oxidative damage during recovery.

Conclusion

By understanding the complex anatomy of the groin and the physiological factors that contribute to the development of inguinal hernias, it is possible to recognise potential risks and adopt effective preventive measures. An adequate diet, rich in essential nutrients, can play a crucial role in preventing inguinal hernias by supporting muscle health and strengthening the abdominal wall. Understanding these elements is essential to minimise the risk of hernia and to promote effective recovery after hernia surgery.

CHAPTER 2

BIOINORGANIC CHEMISTRY AND HEALING

The Role of Metals in Tissue Healing

Metals play distinct but complementary roles in tissue healing, acting as enzymatic cofactors, regulators of the immune response and essential elements in the structure of proteins. Here is an in-depth analysis of the specific contribution of each metal:

Zinc :

Enzyme cofactor: Zinc is an essential cofactor for many enzymes involved in tissue repair, in particular matrix metalloproteinases (MMPs). These enzymes break down extracellular matrix (ECM) components such as collagen, thereby facilitating tissue remodelling.

Collagen synthesis: Zinc is crucial for collagen synthesis via activation of enzymes from the collagenase family. It also plays a part in regulating the genes involved in collagen production, encouraging the formation of new collagen fibres which are essential for healing.

Copper :

Angiogenesis: Copper is essential for angiogenesis, the process of forming new blood vessels. It activates angiogenic growth factors such as vascular endothelial growth factor (VEGF) and tyrosinase, facilitating the vascularisation of regenerating tissue. Antioxidant : Copper acts as a cofactor for superoxide dismutase (SOD), an enzyme that neutralises superoxide radicals, protecting cells against oxidative damage and reducing oxidative stress during tissue repair.

Iron :

Haemoglobin: Iron is an essential component of haemoglobin, which transports oxygen to damaged tissue. Adequate oxygenation is crucial for cell survival and tissue repair. Inflammatory response: Iron is involved in modulating the inflammatory response by regulating the production of cytokines and cell adhesion molecules, thereby controlling cell proliferation and immune responses during wound healing.

Manganese :

Bone formation : Manganese is essential for the biosynthesis of glycosaminoglycans (GAGs) and proteoglycans, major components of cartilage and bone. It acts as a cofactor for enzymes involved in bone matrix formation, facilitating the regeneration of connective tissue. Antioxidant : Manganese is a cofactor of manganese superoxide dismutase (MnSOD), an antioxidant enzyme that protects cell mitochondria from oxidative damage, supporting wound healing.

Selenium :

Antioxidant : Selenium is a component of glutathione peroxidase (GPx), an antioxidant enzyme that reduces peroxides, limiting oxidative damage to cell membranes and proteins during tissue regeneration.

Modulation of inflammation: Selenium plays a role in modulating immune and inflammatory responses, by influencing the production of cytokines and regulating the activity of immune cells, thereby optimising the healing process.

Magnesium :

Cellular stability: Magnesium contributes to the stability of cell membranes by regulating ion channels and intracellular signals. This promotes efficient cell communication and supports the regeneration of damaged tissue.

Muscle relaxation: Magnesium plays a part in muscle relaxation by acting as a natural calcium antagonist, reducing muscle spasms and facilitating muscle recovery after injury.

By interacting with specific enzymes and regulating critical biological processes, these metals collectively contribute to tissue healing. Understanding their role and their complex interactions makes it possible to design optimised therapies to promote recovery after inguinal hernia or similar injuries.

Interaction of Metals and Nutrients in Tissue Healing

Trace elements and metalloproteins play a vital role in the regeneration of muscular and connective tissues, which are essential for the repair of inguinal hernias. Trace elements such as zinc, copper and iron act as enzymatic co-factors in the metabolic processes involved in protein synthesis and regulating cell growth.

· Zinc: Essential for cell proliferation and collagen synthesis. As a cofactor for enzymes of the matrix metalloproteinase (MMP) family and collagenases, zinc facilitates remodelling of the extracellular matrix. It also stabilises the structure of DNA and RNA, facilitating the gene transcription required for tissue repair. In addition, zinc influences the function of immune cells, modulating inflammatory responses and the release of pro-inflammatory cytokines, which are essential for balanced healing.

· Copper: Plays a vital role in angiogenesis, the process of forming new blood vessels, crucial for the revascularisation of damaged tissue. As a cofactor for the enzyme lysyl oxidase, copper catalyses the cross-linking of collagen and elastin fibres, reinforcing the structure of new tissue. It also influences the migration and proliferation of endothelial cells, improving the supply of nutrients and oxygen required for wound healing.

- Iron: Essential for oxygen transport and cellular energy production, it is a key component of haemoglobin and myoglobin, which provide oxygen to repair tissues. Iron is involved i n reduction-oxidation reactions within mitochondria, promoting the production o f ATP, which is necessary for the proliferation and migration of repair cells. It is also a component of certain antioxidant enzymes, such as catalase, which protects cells from oxidative damage during healing.

Metalloproteins, such as metallothionein, also play an important role in regulating the homeostasis of metal ions such as zinc and copper. They influence inflammatory processes and antioxidant defence mechanisms, modulating the release of these ions in response to cellular signals, affecting immune cell migration, cell proliferation and collagen synthesis.

By understanding the importance of these interactions between metals and nutrients, it becomes possible to develop targeted therapeutic approaches, such as optimised nutritional supplements or biomaterials impregnated with specific metal ions, to promote the healing of inguinal hernias. Advances in bioinorganic chemistry are shedding valuable light on the role of metals in tissue healing, making it possible to design strategies that are more effective than ever before. innovative therapies designed to optimise the repair of tissues affected by inguinal hernias and other similar lesions.

Impact of Trace Elements and Metalloproteins on Tissue Repair

In addition to their role as enzyme cofactors, trace elements and metalloproteins interact in complex ways with other cellular components to orchestrate tissue repair. For example, manganese is another crucial trace element that acts as a cofactor for superoxide dismutase (SOD), an antioxidant enzyme that protects cells from oxidative damage during the inflammatory phase of wound healing. Manganese-dependent SOD neutralises superoxide free radicals, reducing oxidative stress and facilitating an environment conducive to tissue repair.Although often overlooked, magnesium plays an important role in stabilising DNA and RNA structures, as well as in energy metabolism via its involvement in ATP synthase. It is crucial for the migration and proliferation of fibroblasts, the cells responsible for producing collagen and the extracellular matrix. In addition to metallothionein, metalloproteins also include proteins such as ferritins and ceruloplasmins. Ferritins store and release iron in a controlled manner, preventing cellular damage induced by free iron, while ensuring a constant supply of iron for metabolic needs. Ceruloplasmin, a copper transport protein, plays a role in iron oxidation, facilitating its incorporation into haemoglobin and myoglobin, and is also involved in antioxidant defences.

Synergistic interactions between different metals and nutrients are also of great importance. For example, copper and zinc are both cofactors in copper-zinc superoxide dismutase (Cu/Zn SOD), which neutralises superoxide free radicals, thereby reducing inflammation and cell damage. Furthermore, an optimal balance between these metals is essential to avoid antagonistic effects where an overload of one can lead to a relative deficiency of the other, thus disrupting the processes ofAdvances in bioinorganic chemistry and biotechnology are enabling the design of advanced biomaterials that can controllably release specific metal ions at the site of injury. These materials can be designed to mimic the properties of natural tissues, improving tissue integration and accelerating repair. For example, zinc-loaded hydrogels or copper-impregnated collagen matrices can be used to directly target the metabolic pathways involved in inguinal hernia healing.

Synthesis and Therapeutic Applications

Understanding the specific roles of metals and their interactions enables innovative therapeutic strategies to be developed to optimise tissue healing. These approaches may include :

· Nutritional supplements: Optimised to include essential trace elements such as zinc, copper, iron and manganese, these supplements can help correct deficiencies and support the biological processes necessary for tissue repair.

· Metal-impregnated biomaterials: Metal-impregnated biomaterials, such as hydrogels or collagen matrices, are designed to progressively release specific metal ions at the site of injury. This controlled release of metal ions promotes key biological processes, such as cell proliferation and extracellular matrix synthesis, which are essential for tissue healing and regeneration.

• Bioinorganic drugs: Advances in bioinorganic chemistry can lead to the design of new drugs capable of precisely modulating biochemical pathways to improve tissue repair. For example, transition metal complexes such as zinc or copper can be used to specifically catalyse the activity of metalloproteases, such as MMP-1 (collagenase) and

MMP-2 (gelatinase), which are involved in the controlled degradation of the extracellular matrix. These drugs can accelerate the degradation of damaged tissue and promote remodelling of the extracellular matrix, which is crucial in the repair of inguinal hernias.

In summary, metals play a multifunctional role in tissue healing. As enzymatic cofactors, immune response regulators and structural components, they are essential to a variety of biological processes. necessary for tissue repair. Trace elements such as zinc, copper, iron, manganese, selenium and magnesium, as well as metalloproteins, contribute to collagen synthesis, the formation of new blood vessels, modulation of the inflammatory response and protection against oxidative damage. Advances in bioinorganic chemistry offer enormous potential for developing innovative therapies aimed at improving tissue repair, particularly in the context of inguinal hernias and other similar injuries.

CHAPTER 3

ROLE OF ACTIN IN TISSUE REGENERATION AND HEALING OF INGUINAL HERNIAS

Introduction to Actin and its Importance

Actin is a ubiquitous and essential protein in eukaryotic organisms. Highly conserved, it is found in virtually all cell types, playing a crucial role in many cellular and physiological processes. It is produced by genes encoding specific actin isoforms, which are differentially expressed according to cell type and biological function. Origin of actin: In mammals, there are several isoforms of actin, including alpha, beta, gamma, delta, etc., which are expressed in specific ways depending on the type of tissue and stage of development.

Actin functions :

1. Muscle contraction: Actin is a major component of the thin filaments in striated muscle (skeletal and cardiac). It interacts with myosin to generate the force required for muscle contraction, thereby

influencing the strength and mobility of muscle tissue.

2. Regulation of cell motility: Outside muscle, actin is crucial for cell motility. It enables cells to move during processes such as cell migration, cell division and the formation of cell extensions such as pseudopodia and lamellipodia.

3. Cellular structure: Actin forms the cytoskeleton of cells, supporting intracellular transport functions and maintaining cellular integrity.

4. Cell signalling: Actin plays a role in regulating cell signalling by acting as a platform for various protein complexes involved in intra- and intercellular signalling.

Actin production and regulation: Actin is produced in the cell cytoplasm in the form of monomers called G-actin. These monomers assemble to form filaments called F-actin, which are essential for its biological functions.

Actin polymerisation is a dynamic process regulated by a variety of binding and regulatory proteins, as well as by complex cellular mechanisms. This process is crucial for cell motility, muscle contraction and the cellular response to extracellular stimuli.

In short, actin is a fundamental protein for many cellular and physiological functions. Its production, regulation and function are closely linked to the health and proper functioning of cells and tissues throughout the body.

In this chapter, we will explore in detail the role of actin in tissue regeneration and healing of inguinal hernias, focusing on the clinical and therapeutic implications of actin alterations in this condition.

1. the role of Actin in muscle contraction

Actin is a fundamental component of the thin filaments of striated muscle, where it forms a helical structure with other associated proteins, such as tropomyosin and troponin. This structure, known as the actin filament, interacts closely with myosin, the main protein of the thick filaments. During muscle contraction, myosin binds to actin, causing relative slippage between the thin and thick filaments, shortening the length of the sarcomere and producing muscle contraction.

The interaction between actin and myosin is regulated by calcium-dependent signals through the troponin-tropomyosin complex, enabling rapid, coordinated contraction of muscle fibres. The dynamics of this interaction are essential for determining the strength and speed of muscle contraction, as well as the mobility of muscle tissue in the body.

2. Actin Modification in Pathological Conditions

In the context of inguinal hernias and other pathological conditions affecting muscle and connective tissue, actin can undergo various alterations. These alterations may include post-translational modifications such as phosphorylation, ubiquitination or enzymatic cleavage, affecting the structure and function of actin in muscle cells.

The article by Higashi-Fujime et al (year) specifically explores the light digestion o f skeletal muscle actin by proteinase K, an enzyme that specifically cuts the peptide bond between methionine and glycine, producing a 35 kDa C-terminal fragment known as proK-F-actin. This fragment, although capable of slowly polymerising into F-actin in the presence of phalloidin, shows a reduced capacity for polymerisation compared with intact actin. Despite this reduction, proK-F-actin retains an overall structure similar to that of normal actin, with a double-stranded organisation typical of actin filaments.

Impact of Modified Actin on Healing of Inguinal Hernias

1. Tissue regeneration

Alterations in actin, such as those observed with proK-F-actin, can have a significant impact on tissue regeneration after inguinal hernia. Altered actin compromises scar tissue formation due to its reduced ability to polymerise into functional actin filaments. These filaments are essential for muscle contraction and mobility, as well as for maintaining the mechanical structure of tissues. Altered actin could lead to the formation of weakened scar tissue, potentially increasing the risk of hernia recurrence.

2. Interaction with Myosins and Muscle Motility

The study by Higashi-Fujime et al. shows that although proK-F-actin has low ATPase activity with myosin, it retains sufficient structure to interact with it. This altered interaction may negatively influence muscle motility and contraction of abdominal muscles. after inguinal hernia. The reduced ability of proK-F-actin to form functional actin filaments may limit the strength and efficiency of muscle contraction, thus affecting functional recovery after surgery.

Strategies to Increase Actin and Improve Healing

1. Potential therapies

To optimise tissue regeneration after inguinal hernia, pharmacological approaches may be considered to modulate actin expression or polymerisation. Specific drugs and growth factors could be used to stimulate the production of functional actin and promote the formation of robust scar tissue. This could include the use of actin polymerisation promoters or regulators of cell signalling involved in tissue regeneration.

2. Exercise and muscle strengthening

Exercise and muscle strengthening can play a crucial role in stimulating actin polymerisation and improving the strength and resilience of abdominal wall muscle tissue. Specific exercise programmes can increase actin expression and functionality, leading to faster and more effective recovery from inguinal hernia. This approach restores normal muscle function and minimises long-term complications.

By developing these strategies, it is possible to significantly improve outcomes for patients with inguinal hernias, by promoting optimal tissue regeneration and minimising the risk of recurrence.

CONCLUSION

In conclusion, this book has explored in depth the complex mechanisms of inguinal hernia through the lenses of bioinorganic chemistry and molecular biology. We have examined the detailed anatomy of the abdominal wall, the complex interactions of metals and nutrients in tissue healing, and the crucial importance of trace elements and metalloproteins in muscle and connective tissue regeneration. In addition, we discussed the latest scientific advances in the field, including the potential impact of gamma radiation on the adrenal glands.This book aims to provide a comprehensive resource for researchers, clinicians and students interested in these interdisciplinary areas. I hope that this exploration of the biochemical and molecular aspects of inguinal hernia has not only enriched your understanding, but also stimulated your interest in future developments in this crucial area of medicine. Thank you very much for reading and for your interest.

GLOSSARY

Canal Inguinal :

· Structure: Oblique passage through the lower abdominal wall, containing the spermatic cord in men and the round ligament in women.

· Location: Extends from the deep inguinal ring to the superficial inguinal ring.

Deep Inguinal Ring :

· Structure: Opening in the fascia transversalis, marking the entrance to the inguinal canal.

Superficial Inguinal Ring :

· Structure : Opening in the fascia of the external oblique muscle, marking the exit of the inguinal canal.

Internal Oblique and Transverse Muscle of the Abdomen:

· Structure: Forms the floor of the inguinal canal, reinforces the abdominal wall.

· Role: Contributes to the resistance of the abdominal wall against internal pressure.

Inguinal ligament :

· Structure: Fibrous band from the anterior superior iliac spine to the pubic tubercle.

· Role: Point of attachment of the groin muscles, forms part of the floor of the inguinal canal.

· Angiogenesis: the process of forming new blood vessels from pre-existing ones, essential for the revascularisation of damaged tissue.

· Biomaterials: Materials designed to interact with biological systems for medical purposes, such as tissue repair or controlled drug delivery.

· Collagen: Major structural protein in connective tissue, essential for tissue strength and integrity.

· Enzyme cofactor: Non-protein substance which helps an enzyme to catalyse a biochemical reaction.

· Haemoglobin: Protein found in red blood cells, responsible for transporting oxygen in the blood.

· Manganese: Trace element required for bone and cartilage formation and involved in antioxidant enzymes.

· Magnesium: Mineral essential for the stability of cell membranes and energy metabolism.

· Metalloproteases: Enzymes that break down proteins in the extracellular matrix, playing a role in tissue remodelling.

· Metalloproteins: Proteins that contain one or more metal ions, essential for their biological function.

· Trace elements : Minerals required in small quantities for the proper functioning of biological processes.

· Cell proliferation: Process by which cells multiply and increase in number.

· Inflammatory reaction: immune system response to injury or infection, characterised by the production of pro-inflammatory molecules.

· Cross-linking: The process of forming covalent bonds between chains of molecules, thereby reinforcing structures such as collagen and elastin.

· Superoxide Dismutase (SOD): Antioxidant enzyme that protects cells from damage caused by free radicals.

· Zinc: Trace element essential for the functioning of numerous enzymes and the synthesis of collagen.

Actin: A protein essential for muscle contraction and the regulation of cell motility, a major component of the thin filaments in striated muscle.

Inguinal hernia: Protrusion of part of the intestine into the inguinal region through a weak point in the abdominal wall.

F-actin: Polymerised form of actin, essential for muscle contraction.

ProK-F-actin: actin fragment produced by light digestion of actin by proteinase K, with a reduced capacity for polymerisation into F-actin.

Phalloidin: Toxin which stabilises F-actin, promoting its polymerisation. Glossary

Angiogenesis: Process by which new blood vessels form from pre-existing vessels, crucial for supplying nutrients and oxygen to tissues undergoing repair.

Enzyme cofactor: Non-protein substance required for the activity of an enzyme. Metals such as zinc, copper and iron can be used as enzyme cofactors.

Collagen: A major structural protein in connective tissue, essential for the strength and integrity of skin, bones, tendons and ligaments.

Cytokines: Small proteins released by cells which have a specific effect on interactions and communications between cells, playing a key role in modulating the immune and inflammatory response.

Glycosaminoglycans (GAGs) : Complex polysaccharides present in the extracellular matrix, important for the structure and function of connective tissue.

Glutathione Peroxidase (GPx):Selenium-containing antioxidant enzyme which reduces peroxides and protects cells against oxidative damage.

Haemoglobin: Protein in red blood cells that transports oxygen from the lungs to the tissues and carbon dioxide from the tissues to the lungs.

Manganese Superoxide Dismutase (MnSOD): Manganese-dependent antioxidant enzyme which protects cells by neutralising superoxide radicals in the mitochondria.

Extracellular matrix (ECM) : A complex network of macromolecules, such as collagen, elastins and glycoproteins, which provide structural and biochemical support to surrounding cells.

Matrix Metalloproteinases (MMPs): A family of enzymes that degrade extracellular matrix components, playing a key role in tissue remodelling and wound healing.

Proteoglycans: Glycoproteins present in the extracellular matrix, which interact with glycosaminoglycans to form a support structure for cells.

Free radicals: Unstable atoms or molecules that have an unpaired electron, making them highly reactive and capable of causing cellular damage.

Inflammatory response: complex reaction of the body to an injury or infection, involving immune cells, blood vessels and molecular mediators, aimed at eliminating the pathogen and initiating tissue repair.

Superoxide Dismutase (SOD): Antioxidant enzyme which catalyses the dismutation of superoxide radicals into oxygen and hydrogen peroxide, reducing oxidative damage.

Tyrosinase: Copper-containing enzyme involved in melanin

biosynthesis and angiogenic processes, influencing skin pigmentation and the formation of new blood vessels.

Vascularisation: Formation of blood vessels in a tissue, enabling the supply of nutrients and oxygen essential for healing and tissue regeneration.

VEGF (Vascular Endothelial Growth Factor) : Signal protein that stimulates the formation of new blood vessels, playing a key role in angiogenesis.

REFERENCES

1- Netter, F. H. (2019). Atlas of human anatomy. Elsevier Masson.

2- Standring, S. (Ed.). (2016). Gray's anatomy: The anatomical basis of clinical practice (41st ed.). Elsevier.

3- Fitzgibbons, R. J., Jr, Forse, R. A., & Groin Hernia Task Force (2018). Update of guidelines for laparoscopic treatment of ventral and incisional abdominal wall hernias (International Endohernia Society [IEHS])-Part B. Surgical Endoscopy, 32(1), 293-311. https://doi.org/10.1007/s00464-017-5803-0

4- Kingsnorth, A., & LeBlanc, K. (2003). Hernias: Inguinal and incisional. Springer.

5- Amid, P. K. (1996). The Lichtenstein open "tension-free" hernioplasty. American Journal of Surgery, 171(3), 193-197.

6- Alberts, B., Johnson, A., Lewis, J., Raff, M., Roberts, K., & Walter, P. (2002). Molecular Biology of the Cell (4th ed.). Garland Science.

7- Barinaga, M. (1994). New clues to how wounds heal. Science, 264(5161), 1663- 1665.

8- Brem, H., & Tomic-Canic, M. (2007). Cellular and molecular basis of wound healing in diabetes. Journal of Clinical Investigation, 117(5), 1219-1222.

9- Burke, J. F., & Yannas, I. V. (1978). Successful use of a physiologically acceptable artificial skin in the treatment of extensive burn injury. Annals of Surgery, 187(4), 379-391.

10- Cohen, M. M. (2006). The role of zinc in the healing of tissue wounds. Journal of Clinical Pathology, 59(4), 321-323.

11- Davis, G. E., & Senger, D. R. (2005). Endothelial extracellular matrix: Biosynthesis, remodeling, and functions during vascular morphogenesis and neovessel stabilization. Circulation Research, 97(11), 1093-1107.

12- Furie, B., & Furie, B. C. (1988). The molecular basis of blood coagulation. Cell, 53(4), 505-518.

13- Kiefbik, A., & Biernat, J. (2019). The role of microelements in the wound healing process. Journal of Trace Elements in Medicine and Biology, 56, 50-56.

14- Laurent, G. J. (1987). Dynamic state of collagen: Pathways of collagen degradation in vivo and their possible role in regulation of collagen mass. American Journal of Physiology, 252(1), C1-C9.

15- Martin, P. (1997). Wound healing--aiming for perfect skin regeneration. Science, 276(5309), 75-81.

16- Prockop, D. J., & Kivirikko, K. I. (1995). Collagens: Molecular biology, diseases, and potentials for therapy. Annual Review of Biochemistry, 64, 403- 434.

17- Weiss, S. J. (1989). Tissue destruction by neutrophils. New England Journal of Medicine, 320(6), 365-376.

18- Higashi-Fujime, S., Suzuki, M., Titani, K., & Hozumi, T. (1992). Muscle actin cleaved by proteinase K: Its polymerization and in vitro motility. Journal of Biochemistry, 112(4), 568-572. 19-Alberts, B., Johnson, A., Lewis, J., Raff, M., Roberts, K., & Walter, P. (2002). Molecular Biology of the Cell (4th ed.). Garland Science.

20-Dos Remedios, C. G., & Chhabra, D. (2003). Actin-binding Proteins. Springer. 21-Pollard, T. D., & Earnshaw, W. C. (2002). Cell Biology (1st ed.). Saunders.

22- Dominguez, R., & Holmes, K. C. (2011). Actin Structure and Function. Springer.

23- Pardee, J. D., & Spudich, J. A. (1982). Purification of muscle actin. Methods in Enzymology, 85(Pt B), 164-181. doi:10.1016/S0076-6879(82)85017-1

24- Cooper, J. A., & Pollard, T. D. (1982). Methods to measure actin treadmilling rate in living cells. Methods in Enzymology, 85(Pt B), 182-210.

25- Selden, S. C., & Pollard, T. D. (1983). Phalloidin and the stabilization of actin filaments. Journal of Biological Chemistry, 258(16), 14461-14466. PMID: 6627651.

26- Higashi-Fujime, S., Suzuki, M., Titani, K., & Hozumi, T. (1992). Muscle actin cleaved by proteinase K: its polymerization and in vitro motility. Journal of Biochemistry, 111(6), 703-708. doi:10.1093/oxfordjournals.jbchem.a123940

Pollard, T. D., & Korn, E. D. (1973). Acanthamoeba myosin. II. Interaction with actin and with a new cofactor protein required for actin activation of Mg2+- adenosine triphosphatase activity. Journal of Biological Chemistry, 248(18), 6375-6380. PMID: 4743355

27- Burgess, D. R., & Chang, F. (2005). Site-directed mutagenesis of the actin gene. Methods in Molecular Biology (Clifton, N.J.), 300, 17-28. doi:10.1385/1- 59259-895-9:017.

Printed by Books on Demand GmbH, Norderstedt / Germany